T4-ADB-939

MAR -- 2022

Super Sea Creatures

SEA CUCUMBERS

GAIL TERP

Black Rabbit Books

Bolt is published by Black Rabbit Books
P.O. Box 3263, Mankato, Minnesota, 56002.
www.blackrabbitbooks.com
Copyright © 2021 Black Rabbit Books

Marysa Storm, editor; Grant Gould, designer;
Omay Ayres, photo researcher

All rights reserved. No part of this book may be reproduced, stored in a retrieval system or transmitted in any form or by any means, electronic, mechanical, photocopying, recording, or otherwise, without written permission from the publisher.

Names: Terp, Gail, 1951- author.
Title: Sea cucumbers / by Gail Terp.
Description: Mankato, Minnesota : Black Rabbit Books, [2021] | Series: Bolt. super sea creatures | Includes bibliographical references and index. | Audience: Ages 8-12 | Audience: Grades 4-6 | Summary: "Dive into learning about sea cucumbers with vibrant photography, engaging text, and range maps, size comparisons, and other infographics"– Provided by publisher. Identifiers: LCCN 2019026794 (print) | LCCN 2019026795 (ebook) | ISBN 9781623102845 (hardcover) | ISBN 9781644663806 (paperback) | ISBN 9781623103781 (ebook)
Subjects: LCSH: Sea cucumbers–Juvenile literature.
Classification: LCC QL384.H7 T47 2021 (print) | LCC QL384.H7 (ebook) | DDC 593.9/6-dc23
LC record available at https://lccn.loc.gov/2019026794
LC ebook record available at https://lccn.loc.gov/2019026795

Printed in the United States. 2/20

Image Credits

Alamy: imageBROKER, 31; Poelzer Wolfgang, 4–5; WaterFrame, 22; Dreamstime: Alexander Ogurtsov, 1, 16–17; Minden: Georgette Douwma, 28 (top); Jurgen Freund, 28 (btm); Shinji Kusano, 12; Science Source: ANT Photo Library, 15 (deep water); D.P. Wilson/FLPA, 21 (top); Franco Banfi, 10–11; Frans Lanting/MINT Images, 15 (shallow water); Jeff Rotman, 9; Martin Shields, 27; Shutterstock: Ethan Daniels, Cover, 32; F16-ISO100, 26–27 (toothpaste); HelloRF Zcool, 14 (cucumber); Krittiya Siriwal, 15 (coral reefs); Laura Dinraths, 6, 21 (btm); Lydia Vero, 18; Macrovector, 24 (plankton); MYP Studio, 14 (bkgd); Nick Kashenko, 24 (sea cucumber); nuruddean, 26–27 (shampoo); photka, 24 (sand); Prachaya Roekdeethaweesab, 24 (crab); Rich Carey, 22–23, 24 (sea turtle); tubuceo, 3; Wolna, 24 (fish)

Every effort has been made to contact copyright holders for material reproduced in this book. Any omissions will be rectified in subsequent printings if notice is given to the publisher.

CONTENTS

CHAPTER 1
In the Sea........................4

CHAPTER 2
Food to Eat
and a Place to Live.......13

CHAPTER 3
Family Life....................19

CHAPTER 4
Part of Their World......25

Other Resources...........30

CHAPTER 1

In the Sea

A sea cucumber searches the ocean floor with its tentacles. As the sea cucumber eats, a fish comes near. **Predator**! The sea cucumber has no time to hide. So it shoots its guts out of its **anus**. Surprised, the fish swims away. The sea cucumber then digs into the sand. It needs to hide so its guts can grow back.

5

COMPARING LENGTHS

spotted worm sea cucumber
black sea cucumber
yellow sea cucumber

6　　　　inches　　　　9

A Good Name

Sea cucumbers are amazing ocean animals. There are many types of them. Most are shaped like cucumbers. That's where their name comes from. They come in many colors. Some sea cucumbers are very short. Others are quite long.

A sea cucumber has no eyes or nose. It just has a mouth. On the other end of its body is its anus. A sea cucumber uses its anus for two main jobs. It sends out waste through it. It also breathes through it.

about 72 inches (183 centimeters)

about 14 inches (36 cm)

about 3 inches (8 cm)

18   27   36   45   54   63   72

Getting Around

Sea cucumbers mostly stay on the sea floor. Many can't even swim. They use their five rows of many feet to get around. The rows go along the bottoms of their bodies. With them, sea cucumbers walk slowly on the ocean floor.

To move to new places, some sea cucumbers fill their bodies with water. They then float toward the surface and ride a **current**.

SEA CUCUMBER FEATURES

MOUTH

10

ANUS

FEET

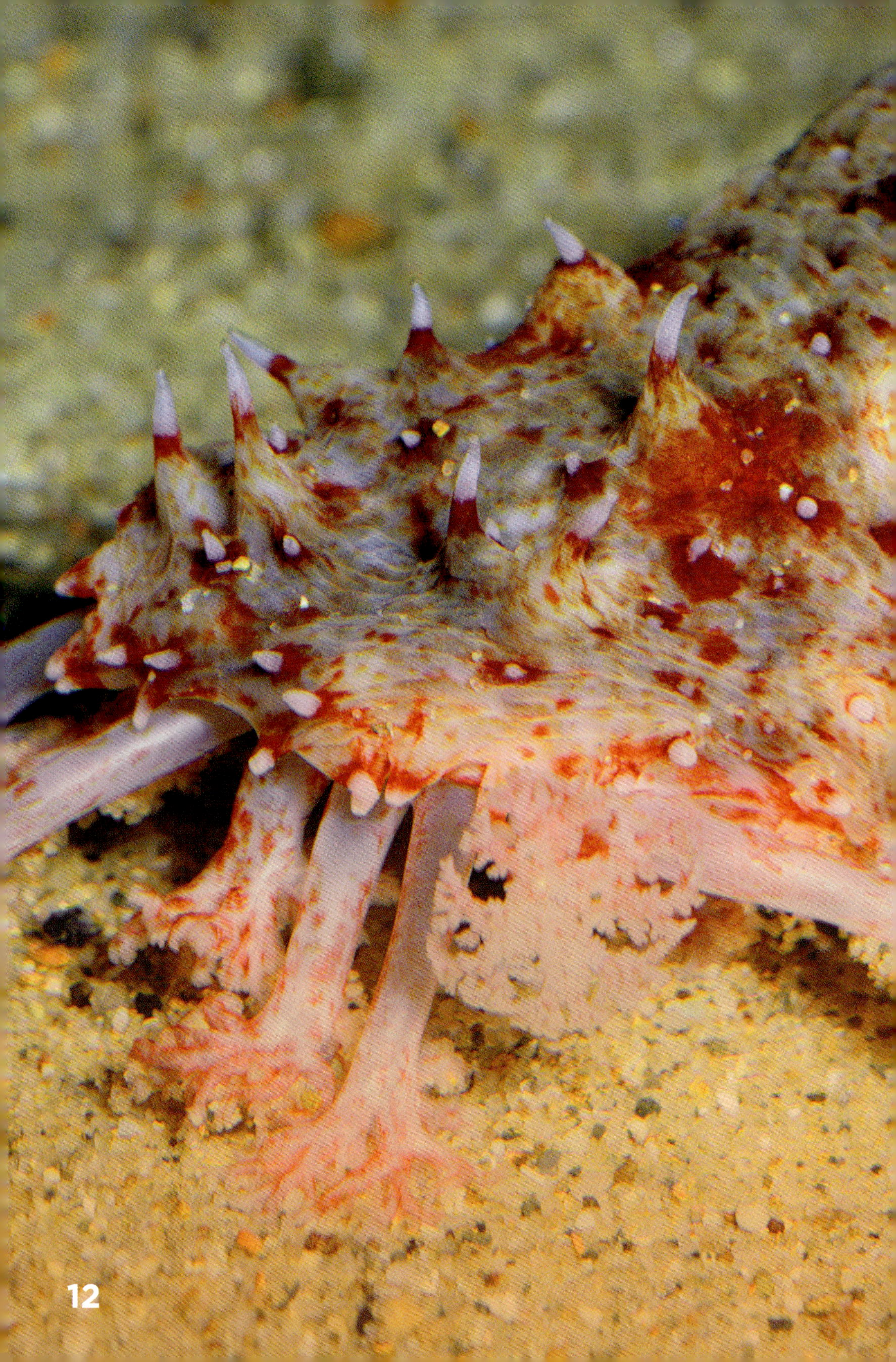

CHAPTER 2
Food to Eat and a Place to Live

Sea cucumbers eat what they can find. They mostly nibble sand containing bits of plant and animal matter. Some also eat **plankton**. They gather food with the tentacles around their mouths.

At Home in the Ocean

Sea cucumbers live in all of the world's oceans. They live in both shallow and deep water. Some spend their lives in deep, cold **trenches**. Others live on coral reefs. Wherever they live, sea cucumbers mostly stay on the ocean bottom.

TRENCHES

Sea Cucumber Homes

CORAL REEFS

DEEP WATER

SHALLOW WATER

15

WHERE SEA CUCUMBERS LIVE

Sea Cucumber Range Map

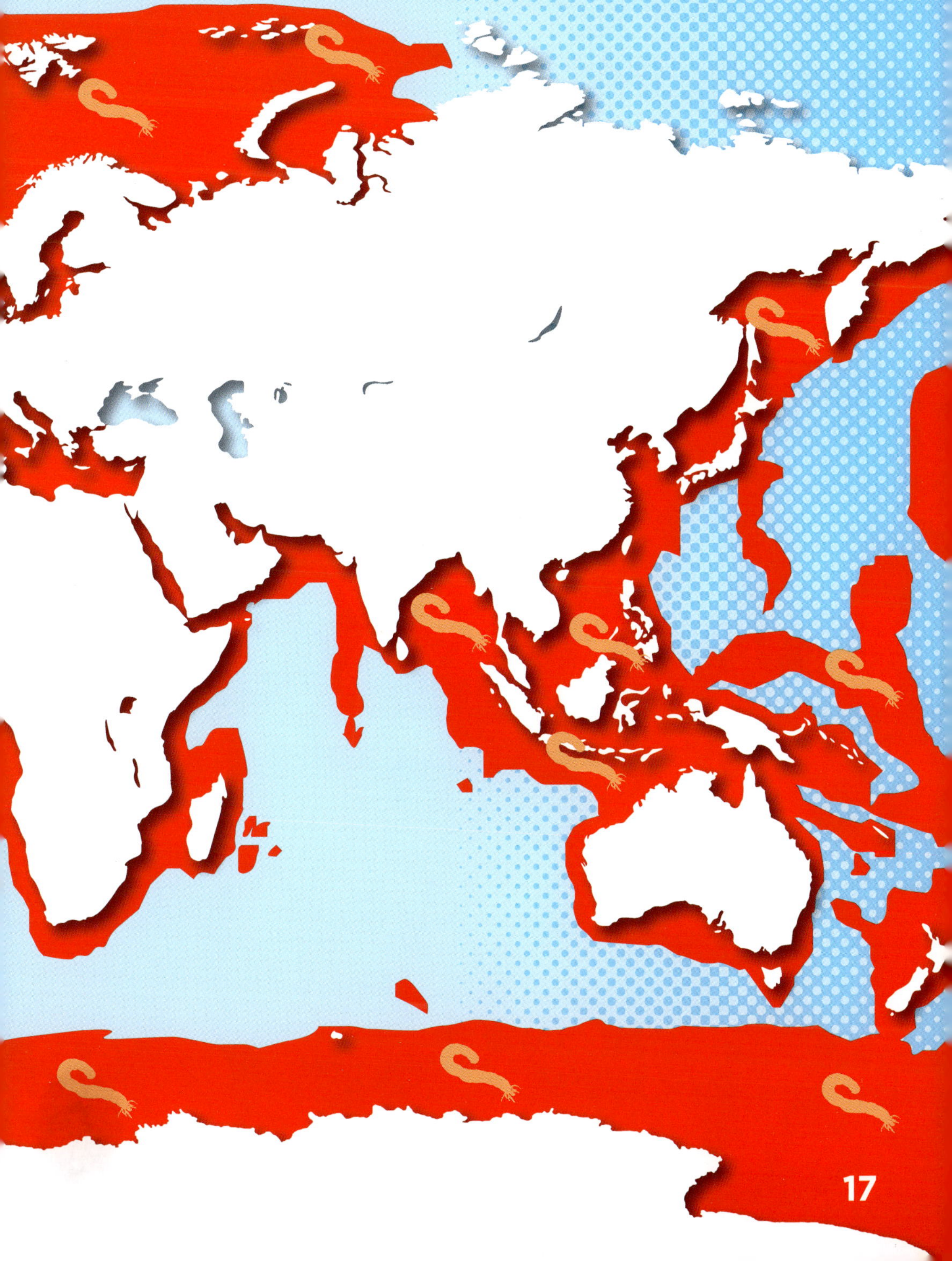

18

CHAPTER 3

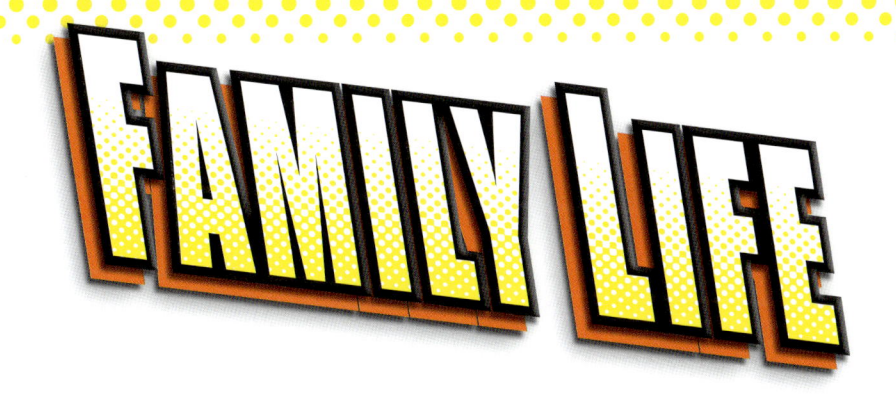

FAMILY LIFE

Sea cucumbers often live alone. When males and females meet, they sometimes **spawn**. The females lay eggs in the water. The males then **fertilize** the eggs. Both parents then leave the eggs.

Growing Up

Sea cucumber eggs hatch into **larvae**. For a few weeks, the larvae simply float in the ocean. Then they sink to the ocean floor. The larvae start to grow into adults. Their slow-walking life has begun.

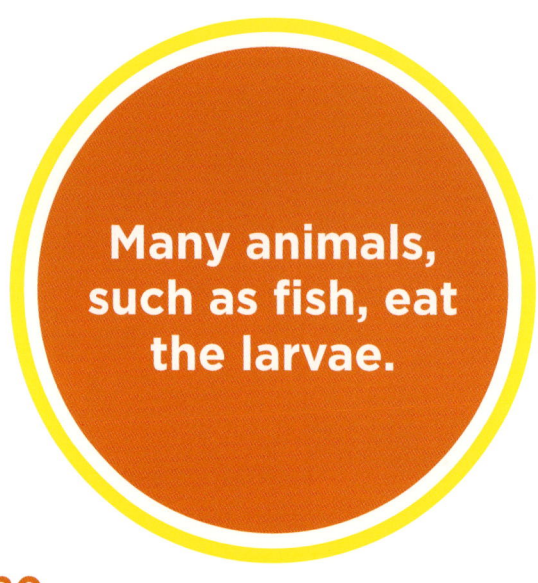

Many animals, such as fish, eat the larvae.

21

8 TO 30
NUMBER OF TENTACLES

BY THE NUMBERS

5 to 10 years
LIFE SPAN

ABOUT 1,200 TYPES OF SEA CUCUMBERS

2 weeks to a few months TIME NEEDED TO REGROW GUTS

Sea Cucumber Food Chain

This food chain shows what eats sea cucumbers. It also shows what sea cucumbers eat.

FISH **CRABS** **SEA TURTLES**

SEA CUCUMBERS

BITS OF PLANTS AND ANIMALS IN SAND **PLANKTON**

24

CHAPTER 4

Part of Their World

Sea cucumbers have predators. Fish and crabs eat them. Sea turtles eat them too. But sea cucumbers have great **defenses**. They can shoot out their guts. Some can shoot out sticky threads too. Predators get tangled in the threads. Then sea cucumbers have a chance to escape. Sea cucumbers can also squeeze into cracks in rocks to hide.

Human Threats

People fish for sea cucumbers. They sell them as food. They sell them for medicines too. But fishing sea cucumbers has caused problems. People are catching too many. Some types might die out.

Some people use sea cucumbers to make shampoo and toothpaste.

27

Sea cucumbers are also helpful for pearlfish. During the day, the small fish like to hide. Some hide in sea cucumbers. The fish get in by swimming up the anus.

Good for the Ocean

Humans need to look after sea cucumbers. They're an important part of the ocean. They're food for many animals. They clean the ocean floor too. The sand they eat comes out as clean waste. This waste helps make the water less **acidic**. Acidic water hurts coral reefs. Sea cucumbers might look strange. But they're very helpful. These animals truly are super sea creatures.

GLOSSARY

acidic (uh-SID-ik)—containing a chemical substance that can dissolve things

anus (EY-nuhs)—the lower opening of the digestive tract

current (KUR-uhnt)—the part of a fluid body, such as air or water, moving continuously in a certain direction

defense (DEE-fens)—a way of resisting attack

fertilize (FUR-tl-ahyz)—to make an egg able to grow and develop

larva (LAR-vuh)—the wormlike form of an animal after hatching from an egg

plankton (PLANK-tun)—the very small animal and plant life in an ocean, lake, or river

predator (PRED-uh-tuhr)—an animal that eats other animals

spawn (SPAWN)—to produce or deposit a large number of eggs

trench (TRENCH)—a long, narrow hole in the ocean floor

LEARN MORE

BOOKS
Lawrence, Ellen. *Poison Slimers: Poison Dart Frogs, Sea Cucumbers & More.* Slime-inators & Other Slippery Tricksters. New York: Bearport Publishing, 2019.

Perkins, Wendy. *Sea Cucumbers.* Weird and Unusual Animals. Mankato, MN: Amicus High Interest, 2018.

Troup, Roxanne. *Deep-Sea Creatures.* Creepy, Kooky Science. New York: Enslow Publishing, 2020.

WEBSITES
Are Sea Cucumbers Vegetables?
oceanservice.noaa.gov/facts/seacuke.html

Sea Cucumbers
www.nationalgeographic.com/animals/invertebrates/group/sea-cucumbers/

Sea Cucumbers
www.nwf.org/Educational-Resources/Wildlife-Guide/Invertebrates/Sea-Cucumbers

INDEX

D
defenses, 4, 23, 25

F
features, 4, 7, 8, 10–11, 13, 22, 28

food, 4, 13, 24, 29

H
habitats, 4, 8, 14–15, 20, 25, 29

human threats, 26

L
life spans, 22

P
predators, 4, 20, 24, 25, 26, 29

R
ranges, 14, 16–17

S
sizes, 6–7

Y
young, 19, 20